Charles Milling

Sense of Place: Its History, Evolution, and Present Application

AF135581

Charles Milling

Sense of Place: Its History, Evolution, and Present Application

LAP LAMBERT Academic Publishing

Impressum / Imprint

Bibliografische Information der Deutschen Nationalbibliothek: Die Deutsche Nationalbibliothek verzeichnet diese Publikation in der Deutschen Nationalbibliografie; detaillierte bibliografische Daten sind im Internet über http://dnb.d-nb.de abrufbar. Alle in diesem Buch genannten Marken und Produktnamen unterliegen warenzeichen-, marken- oder patentrechtlichem Schutz bzw. sind Warenzeichen oder eingetragene Warenzeichen der jeweiligen Inhaber. Die Wiedergabe von Marken, Produktnamen, Gebrauchsnamen, Handelsnamen, Warenbezeichnungen u.s.w. in diesem Werk berechtigt auch ohne besondere Kennzeichnung nicht zu der Annahme, dass solche Namen im Sinne der Warenzeichen- und Markenschutzgesetzgebung als frei zu betrachten wären und daher von jedermann benutzt werden dürften.

Bibliographic information published by the Deutsche Nationalbibliothek: The Deutsche Nationalbibliothek lists this publication in the Deutsche Nationalbibliografie; detailed bibliographic data are available in the Internet at http://dnb.d-nb.de. Any brand names and product names mentioned in this book are subject to trademark, brand or patent protection and are trademarks or registered trademarks of their respective holders. The use of brand names, product names, common names, trade names, product descriptions etc. even without a particular marking in this works is in no way to be construed to mean that such names may be regarded as unrestricted in respect of trademark and brand protection legislation and could thus be used by anyone.

Coverbild / Cover image: www.ingimage.com

Verlag / Publisher:
LAP LAMBERT Academic Publishing
ist ein Imprint der / is a trademark of
OmniScriptum GmbH & Co. KG
Heinrich-Böcking-Str. 6-8, 66121 Saarbrücken, Deutschland / Germany
Email: info@lap-publishing.com

Herstellung: siehe letzte Seite /
Printed at: see last page
ISBN: 978-3-659-56500-7

Copyright © 2014 OmniScriptum GmbH & Co. KG
Alle Rechte vorbehalten. / All rights reserved. Saarbrücken 2014

Table of Contents

Abstract

SENSE OF PLACE: ITS HISTORY, EVOLUTION, AND PRESENT APPLICATION

Charles R. Milling, M.S.

George Mason University

Thesis Director: Dr. Susan Crate

To date, the concept of 'sense of place' has/is used in several disciplines and interdisciplinary applications. Are these different fields using the term differently and, if so, how? Does the concept prove useful in academic, political and other fields? What is(are) the origin(s) of the concept, how did it evolve in its meaning and uses over time, and how is it being used today? In this thesis my main objective is to disentangle the many meanings of the term to provide others using it in contemporary contexts with an exhaustive and comprehensive review of the term in hopes that this will increase the concept's utility. To these ends, the overarching question of my thesis is, "What are the origin(s), evolution(s) and present-day uses of the concept of 'sense of place'? My primary goal was to provide a historical review and see where the concept originated, either in one or more than one context, how the meanings differ or overlap, and how the uses changed over time and to the present day. Once I have completed this thorough and exhaustive review as described, I will add some ideas about the possibility of a unifying theme. However, this is tangential and speculative at best, seeing as my main effort is tracing the concept's history, evolution and contemporary uses.

Chapter 1

Introduction

Sense of place has been studied from across several disciplines, bringing about several different definitions and conceptualizations(Casey, 1996; Devine-Wright & Clayton, 2010a; Gruenewald, 2003; Hay, 1998a; H. Proshansky, Fabian, & Kaminoff, 1983; Relph, 1976).. The several different definitions and conceptualizations make it difficult for researchers to apply the sense of place, leading some to conclude that the concept stands at a crossroads (Devine-Wright & Clayton, 2010). Some argue that before sense of place can advance the field it needs an integrative meaning (Devine-Wright & Clayton, 2010). In order to determine if there is a common meaning, it is critical to first look at the concept's history and evolution. This will provide insight into whether sense of place can be used to understand what connects people to local places, broadens their ecological awareness, and moves them to lead more sustainable lives.

The literature to date reveals a concept that has been used in a myriad of ways. Though this has helped to provide sense of place with depth, the many uses has made it difficult to understand what it means, and because it is difficult to understand what it means, it becomes difficult for researchers to use the concept in cross-disciplinary ways, hence the roadblock. This study could go a long way in

1

resolving these issues and it is the first step in a longer journey. Place, and how people make connections with places, fascinates me.

For me, this relationship begins with how a person senses place and how they form their sense of place. However, before I can address the *how* and *why* involved with sense of place, we must tackle the *what*. What is sense of place? More importantly, what does it mean given that in the Western world we live in an ever-increasing globalized society.

Study Aims and Research Questions

My study aims to present an historical and contemporary literature review of sense of place. To these ends, my thesis's main research question is: What are the origin(s), evolution(s) and present-day uses of the concept of 'sense of place'? In the process of addressing that question, I will explore sub-questions such as, Who coined the term? How have researchers from different disciplines used sense of place? Is it possible to create a unifying sense of place concept? And, Can it be used in the sustainability movement?

Justification of Literature Review Towards my Research Aims

Sense of place is a popular concept among researchers in several major disciplines today including environmental studies. As society becomes increasingly globalized, some argue that people are losing their connection to place, which works against creating sustainable communities (Escobar, 2001). Society needs to focus back on the local perspective and on local places. This sentiment has certainly taken hold among some segments of the American population. More and more, people are looking towards obtaining resources from local sources. People shop more at local markets and farmers markets. In its second year the Saturday Morning Market, a farmers market in Tampa, Florida, drew about 2,000 people a week, but now they draw over 10,000 (Scullin, 2009). One Concord farmer

2

reported seeing a 40% increase in sales at farmers markets (Flower, 2008). Does this push for more localized economies signal a fundamental change in society's paradigm? Is the U.S. becoming slightly less global? Are people looking to find fulfillment in their local places? This study could go a long way in informing the local perspective. As Stewart (1996) notes "far from being a timeless or out-of-the-way place, the local finds itself reeling in the wake of every move and maneuver of the center of things" (p. 137). Having a sense of place forms the heart of local communities (Hay, 1998a).

Approach

To carry out my study I will conduct an exhaustive literature review of the sense of place concept. The literature review will include a historical review and will assess its various meanings and themes. To these ends, I will employ a thematic analysis, identifying the major theme(s) across the literature (Babbie, Halley, & Zaino 2007). This analysis will allow me to see which meanings might be better in terms of clarity.

Literature Review

A review of the literature reveals a major knowledge gap. The gap deals with the lack of a clear meaning in sense of place and a lack of empirical data. To begin, we should first define sense of place, but pinpointing a definition for sense of place is difficult due to the variety of approaches (Devine-Wright & Clayton, 2010b).

Sense of place is defined in several disciplines, including the behavioral sciences and in geography. In the behavioral sciences, definitions center on the individual's cognitive, affective, and behavioral associations with certain places. For instance, Stedman, (2002) defines sense of place as "the interplay of affect and emotions, knowledge and beliefs, and behaviors and actions in reference to a place" (pg. 5). Low & Altman define sense of place as place attachment or "an individual's cognitive or

3

emotional connection to a particular setting or milieu" (1992: 165). Notice that with Stedman's definition the focus lies in meanings given to a place versus Low & Altman's definition, which puts the focus on an individual's connection to a place. The former definition has more emphasis on places themselves, thus helping to articulate two distinct yet related approaches. These approaches will become more evident in the later chapters. Although these definitions differ in the terminology they use, largely due to disciplinary norms, they both essentially focus on how sense of place is about the emotional and cognitive connection to a place.

Other theorists, beginning with cultural geographers such as Relph and Tuan define sense of place in terms of three components; physical setting, activities within a place, and meanings (Gustafson, 2001a). Early work by cultural geographers argue that local geographies, through the idea of landscape, have special meanings, but these meanings become difficult for individuals to understand (Gustafson, 2001). The early human geographers appear to be the first to examine sense of place as a concept and it was in Relph's (1976) seminal text, *Place and Placelessness*, that I first saw the term sense of place discussed as a concept. Relph (1976), Tuan (1974, 1980), and Buttimer (1980) had major influence on establishing sense of place as a viable research topic, especially for geography and phenomenology. All works in those fields since then reference and draw heavily on their ideas.

Anthropologists hold similar views to Tuan and Relph and posit that that places themselves carry meanings through the collective myths, stories, daily actions, etc. of a community, and these meanings then inform an individual's knowledge, awareness, and values about places (Kahn, 1996). As Kahn, (1996) notes, "places capture the complex emotional, behavioral, and moral relationships between people and their territory" (pg. 168). In this way, places themselves have an identity or a tactile sensate force (Stewart, 1996).

Basso (1996) conceptualizes sense of place using Heidegger (1996) notion of *dwelling*, which assigns importance to the forms of awareness or consciousness with which people perceive and

4

understand geographical space. Dwelling consists in the multiple "lived relationships" that people maintain with places, and these "lived relationships" allow space to acquire meaning and have value in the consciousness of one's being (Basso, 1996). These multiple lived relationship includes features of the physical environment as well as the social environment (Cloke & Jones, 2001). Thus, through dwelling in places, individuals form strong connections with their environments and uses places to help shape their identity. As Basso (1996) writes, "selfhood and placehood are intertwined" (pg. 86).

Casey, (1996) a phenomenologist, further explores this relationship by looking at how the actual living moving body influences places and is influenced by place. Physically people have "lived" bodies that reach out organically into the space surrounding them (Casey, 1996). Several members that exhibit the lived body move with corporeal intentionality. Due to this intentionality, the lived body becomes integrated with its immediate physical surroundings, its concrete place (Casey, 1996). In other words, the space surrounding us isn't empty; it's filled with meaning, value, and substance exhibited by us as lived engaged bodies.

Casey draws on the work of Husserl and Kant in showing how the lived body becomes the natural subject of perception. The subject perceives the immediate place he or she finds themselves and the place extends its own influence on the subject; place integrates with body as much as body integrates with place (Casey, 1996). Basso (1996) calls this process "interanimation." Thus, having a sense of place in this context would logically mean that one has a deep connection to the place. In this vein at whatever level sense of place operates in a person, it goes beyond simply having knowledge of a place.

This merging of selfhood and placehood can come through in several forms such as one's identity, one's behavior in the community, or one's values to the place and to each other (Basso, 1996). The place itself is thought to negotiate social relationships, relationships between the self and the other.

As Kahn (1996) writes, "places capture the complex emotional, behavioral, and moral relationships between people and their territory" (pg. 168).

Sense of place may be thought of as a scaled quality of experience whereby the longer one stays in one place and has experiences in that particular place the stronger sense of place one will have (Relph, 1976). Accordingly, one must interact with a place in order to gain experiences in that place and begin to develop a sense of place. Local places make us who we are; the attributes of place help to shape our identity and possibilities (Gruenewald, 2003). Just as the self and place integrate with each other, so does culture and place; culture and place are intertwined and one cannot be separated without the other (Feld & Basso, 1997).

Sense of place is socially negotiated and encompasses cognitive (knowledge as place meaning) and affective (place attachment; attitudes and preferences as place meaning) dimensions of the self (Semken & Freeman, 2008). Individuals assign/ designate meaning or importance to places over the course of their lives through gaining knowledge, both formal and experiential, of those places and by forming certain attitudes and preferences about those places (Casey, 1996). Place attachment provides an easier epistemology for which to measure the relationship (Devine-Wright & Clayton, 2010a). The above studies in anthropology and geography stem from a phenomenological perspective that emphasizes people's experience in actual places, thus giving places meanings, authenticity, and value as well as helping people better understand themselves (Casey, 1996; Kahn, 1996)

Psychology is another major field understanding how people use experiences in places as a source of their identity. Proshansky (1978), in the late 70's and early 80's, argued that as individuals socialize in different settings they form their identity partly based on the settings of those socializations. Over time individuals can become attached to certain places and place types through positive and negative socializations (Proshansky, Fabian, & Kaminoff, 1983). Proshansky and other environmental psychologists have called this "place attachment." Though that concept is different from sense of place

6

it has many similar qualities. Proshansky's work appears to inform all subsequent studies of sense of place.

Both broad areas of place studies have ties into the sustainability movement in a couple of ways. For one, both sense of place as it has been used as well as the more psychological-type studies on place seek to connect people to actual local places (Hay, 1998; Low & Altman, 1992; Casey, 1996; Cloke & Jones, 2001). Connecting people to local places is echoed in many calls for sustainability.

Some researchers such as Mitchell Thomashow seem to think so. Thomashow (1995) has studied how people interact with places over time and go through different stages of a place-making process. Places become the learning laboratories of how life works, and on the ecology of life. If individuals tap into their sense of place they could become more reflective of how their daily lives affect the environment around them, thus helping them lead more sustainable lifestyles.

Other work by bioregionalists such as Robert Thayer argues for sustainability that is partly based on having a sense of place. Thayer (2003) argues that the more one knows about a place the more likely they will care about that place and the more one cares about the place the more likely they are to take care of it.

Chapter 2

Early Uses

The concept of sense of place is first used in the 1970's, beginning in the field of geography. Early human geographers were curious about how individuals perceived places Seamon (1979). How did people come to understand physical places through their perception of place? How did people become rooted in one place and not another? From where did these sometimes seemingly personal connections with places stem? These questions drove geographers' research.

Though the first usage of the term I found came from a brief review by Prince (1974), he never explored the term in his text but used it as a title to a short review in a journal. It was Edward C. Relph who coined the term and provided an in-depth examination of sense of place. Relph wrote his seminal work, *Place and Placelessness*, for his master's thesis. In it he outlined his views on place and the relationship between people and places and paved the way for other researchers.

Another human geographer, Y.F. Tuan, contributed his seminal work, *Topophilia: A Study of Environmental Perception, Attitudes, and Values*, about this time. Tuan also examined how people perceive places and in doing so become connected to places, but he conceptualized it differently than Relph. Relph and Tuan's work in the mid to late 1970's laid the foundation for all future sense of place

8

studies (Seamon, 2000). Since Relph coined the phrase "sense of place," I will begin with his account.

The crux of Relph's conceptualization of place is the idea of rootedness and authenticity of place. According to Relph (1976) individuals have positive and negative experiences in place(s). Over time, as groups of people form communities and have shared experiences in a place, these places can take on characteristics of their own and form their own senses of place. This creates opportunities for individuals to have an authentic sense of place.

If a person connects with that authenticity in a place, then the person feels "a direct and genuine experience of the entire complex of the identity of places- not mediated and distorted through a series of quite arbitrary social and intellectual fashions about how that experience should be, nor following stereotyped conventions" (Relph, 1976 pg. 64). The more a person feels this in a place and then lives out their daily lives through the context of the "complex identity of the place" the more he or she becomes rooted in a place, thus allowing them to have strong affective ties to a place. The person makes meaning of their own lives through the context of that place. However, for Relph sense of place all starts with an individual sensing a place or having direct sensory experiences with a place (Seamon & Sowers, 2008).

The other main facet of Relph's concept deals with insideness and outsideness, incorporate both positive and negative experiences in places. Having an insideness gives the individual a feeling of safety and security rather than feeling threatened. A person feels comfortable and enclosed rather than exposed and vulnerable, relaxed rather than stressed. On the other hand outsideness gives the individual a feeling of separation and alienation from a place and the community within that place (Relph, 1976). This can lead to a distinct division between the person and the world around them (Seamon, 1996).

This insideness versus outsideness also led Relph to discuss sense of place in terms of strengths of association between the individual and the place. Relph called the greatest strength existential insideness. Existential insideness creates a situation of larger scale, unself-conscious immersion in

9

place (Relph, 1976). Relph argues this is the experience that people typically feel when they are truly "at home" in their own community. In this moment the individual cannot separate themselves from the place. In the subconscious, place and self merge and mirror each other. Ultimately for Relph place and more specifically sense of place is a complex of many things; "By taking place as a multifaceted phenomenon of experience and examining the various properties of place, such as location, landscape, and personal involvement, some assessment can be made of the degree to which these are essential to our experience and sense of place" (pg. 20). A more developed sense of place is based on having important experiences in place over extended periods of time.

Whereas Relph looked at both positive and negative experiences of place, Tuan (1974) concerned himself primarily with the positive aspects of place. His seminal work, *Topophilia: A Study of Environmental Perception, Attitudes, and Values,* examined the affective ties people have with significant places. Like Relph, Tuan's starting point was experience and perception. Though affective bonds form psychologically, they illicit emotional responses, and in the case of topophilia, defined by Tuan as a love of certain places or landscapes, those responses happen positively through love of place or sentimental attachment to local places or landscapes (Tuan, 1974). Tuan wrote, ""more permanent and less easy to express are feelings that one has toward place because it is home, the locus of memories, and the means of gaining a livelihood" (pg. 93).

First and foremost Tuan concerned himself with the sensory perception of places. He examined the major senses of vision, touch, hearing, and smell individually as well as collectively (Tuan, 1974). Through a more collective sensory perception of place one can gain a more dynamic relationship with the places they inhabit (Tuan, 1974). As such, not only do places or environments provide certain identity traits, they also "provide sensory stimuli, which as perceived images lend shape to our joys and ideals" (Tuan, 1974 pg. 113). In other words places give us concrete forms that symbolize, represent, and articulate feelings we have towards many things.

10

Whereas Relph, Tuan, and other early human geographers such as Buttimer (1980) tended to focus on how places take on meaning as the individual gains meaning from a place, psychologists such as Harold Proshansky centered on the individual. Proshansky considered the importance of place in shaping identities as he examined the influence of the *built environment* on urban life, assuming that there is no physical setting that is also not a social, cultural, and psychological (Proshansky, 1978). He argued for an integrated view of place and the self where an individual begins to shape part of their identity based on the socialization dynamic that happens in different places (Proshansky, 1978). He assumed that complex behavior comes from lasting personality traits that can in turn can be influenced or shaped from interactions in different places or settings. These settings enhance or limit the influence of the structure of the personality. Furthermore he argued that human-environment relationships must be a part of a person's personality structure (Proshansky, 1978). In other words, place, through socialization within that place, shapes a person's identity.

Proshansky was one of the first to highlight the physical aspects of place in identity development. For instance, a school represents not just people, including pupils, teachers, and principals; it also becomes a building with classrooms, play areas, and a principal's office (Proshansky, 1978). In another example, the family represents not just mothers, fathers, brothers, and sisters; it also represents a home. This concept can be applied to other aspects of the place-self relationship. As such, since the self-identity is structured by various more specific identities, sometimes referred to as sub-identities, such as social class, occupation, religion, ethnicity, etc., then it follows that there must be a place-identity which becomes subsumed and incorporated into the larger self-identity (Proshansky, 1978).

Proshansky (1978) defined place-identity as "those dimensions of self that define the individual's personal identity in relation to the physical environment by means of a complex pattern of conscious and unconscious ideas, beliefs, preferences, values, feelings, goals, and behavioral tendencies

11

and skills relevant to their environment" (pg. 155). Thus the physical environment helps to shape an individual. Herein lays the potential to also gain insight into an individual's environmental values, attitudes, and behavioral orientations as they relate to specific places. Places themselves become important mediators.

After cultural geographers Relph and Tuan, and psychologist Proshansky, all other sense of place studies and place identity/ attachment studies have conceptualizations similar to either and some both (J.A. Agnew, 1989; Buttimer, 1980; Cuba & Hummon, 1993; Gustafson, 2001a; Hay, 1998a). As Prince (1974) notes, "in the study of the environment the objectives of psychologists and geographers are distinct and complementary. Psychologists are interested in environment in so far as it plays a part in shaping human behavior. Geographers are interested in it for its own sake and the people in it as visitors, inhabitants or landscape makers" (pg. 71). What is important to note is that they opened the door, very widely, for subsequent studies.

Subsequent Uses in the Different Disciplines

A review of sense of place after Relph, Tuan, and Proshansky shows studies in four major disciplines; environmental psychology, phenomenology, human geography, and anthropology. I will now review the use of sense of place in those fields. The literature is vast and here I can only represent some. I chose the fields below because they contained the historical roots and because they contained more studies than in any other fields.

Uses in the Field of Psychology

Following Proshansky are a multitude of studies examining place attachment in psychology (Cuba & Hummon, 1993; Fishwick & Vining, 1992; Jorgenson & Stedman, 2001; Low & Altman, 1992; Proshansky et al., 1983). A search on APA PsycNET database using the key word "place attachment," returned 1,167 results, many of them from the *Journal of Environmental Psychology*. The subjects ranged from studying adults in residential neighborhoods to adolescents in their hometowns and used similar methodology and format. Researchers would survey residents of a community, neighborhood, city, residence, etc. using an ordinal scale and ask them about how attached they feel to that area and in what different areas do those attachments come through. I include here the research most relevant to my purposes.

Korpela (1989) surveyed 9, 12, and 17-year old children on how they self-regulate and form their identity through familiarization and identification with their favorite places. The researcher defined place identity as the emotional attachment to place via places' means of regulating the pleasure/ pain balance and one's self-esteem. According to Korpela (1989) cognitive and affective structures of place identity are the result of an active self-regulation process. Korpela tested this idea with a longitudinal study of children, ages 9, 12, and 17, and found that the majority of respondents did go through self-regulation processes via favorite places. Places helped the children regulate their pleasure and pain (Korpela, 1989).

An often cited work (175 times) in this field came from Jorgenson & Stedman (2001) who examined sense of place as an attitude. The researchers argued that sense of place could achieve greater coherence if viewed as a framework compromising cognitive, affective, and conative processes. Considering sense of place as an attitude rather than a set of deep affective ties, offers many benefits to place researchers such as organization of disorganized constructs, linkage to established literature, and established research methods (Jorgenson & Stedman, 2001). Researchers have an easier time measuring attitudes empirically, and can create behavioral models. Attitudes test better in surveys and studies. They can be rank ordered. So, if sense of place can be talked about using items in a survey then it could potentially be measured and assessed.

To test this idea they surveyed residents across 8 small lakes in Vilas County, Wisconsin. They

13

constructed their sense of place model using the concepts of place attachment, place identity, and place dependency developed by earlier researchers (Williams & Roggenbuck, 1989;Williams, Patterson, Roggenbuck, & Watson, 1992;Moore & Graefe, 1994). Their results indicate that the tripariate structure does create a general sense of place dimension that reflected property owners' thoughts, emotions, and behavioral beliefs. Interestingly they found that the concept of place attachment was more synonymous with sense of place than was place identity or place dependence. Place attachment acted more like sense of place in that it stemmed from affective bonds rather than cognitive bonds. Sense of place as it was originally discussed stemmed from affective bonds. In this study place attachment was seen as primarily an affective construct.

Feldman (1990) expanded the spatial scales involved in place identity to look at the idea of settlement. Given the increasing mobility of individuals in US society, her study explored how individuals identified with place. Though individuals at times strongly identified with "home" places, the question became, what about larger places such as cities and suburbs. Feldman (1990) conceptualized this as "settlement identity." She surveyed residents of Denver, Colorado and found that a majority of those surveyed identified more with the location as a settlement-type rather than the settlement itself. Participants categorized the settlements in which they lived. Participants identified more with living in the suburb versus downtown regardless of the actual suburb in which the individual lived.

Cuba & Hummon (1993) analyzed place identity among three different spatial scales; dwelling, community, and region. They surveyed residents of Cape Cod to see if a person's place identity differs at different scales. The researchers drew on place identity concepts from Proshansky et al. (1983), Altman and Chemers (1980), and Duncan (1982), as well as sense of place concepts and dwelling concepts from Relph (1976) and Buttimer (1980). The researchers found that respondents identified with each of the three levels of place identity at roughly the same frequency with a slightly higher rate

14

going to the dwelling location. Interestingly they found that respondents did not need a longer stay in Cape Cod to identify with the "Cape" as a dwelling. This suggests that length of stay or residency over time does not always play a significant role as previously suggested by many of the human geographers (Cuba & Hummon, 1993).

Yet another noteworthy study came from Hidalgo & Hernandez (2001), which was cited 185 times. They examined place attachment at different spatial scales and the conceptual framework for place attachment. They outlined various researchers' conceptualizations of place attachment and sense of place. For their empirical work they interviewed 177 people from different areas of Spain at different spatial scales. They assessed the house, neighborhood, and city scales. They found three important things; attachment to neighborhood is the weakest, social attachment is stronger than physical attachment to a place, and the degree of attachment varies with age and sex. Place itself it seems isn't as prevalent as the relationships we form in those places.

Overall, the vast majority of sense of place studies in psychology cite Low & Altman's (1992) definition of place attachment as "an individual's cognitive or emotional connection to a particular setting or milieu" (pg. 165). Beyond those, more studies continued in the phenomenological and human geography fields.

Uses in the Field of Phenomenology and Human Geography

The prominent human geographers who explored sense of place after Relph and Tuan include Seamon (1982), Agnew (1989), Shamai (1991), Casey (1996), Hay (1998), and Massey et al. (1999).

Shamai (1991) constructed a scaled sense of place to see if there is a difference in intensity at different scales. Shamai tested three levels of sense of place among Jewish students in Toronto. The levels were metro Toronto, the province of Ontario, and the nation of Canada. Shamai ranked the different levels on an ordinal scale (i.e. strongly agree, somewhat agree, agree, etc.) and then created 6

15

degrees of strength of association with a place. The associations ranged from 0, not having any sense of place, to 6, sacrifice for a place. Shamai writes that a '6' "involves the deepest commitment to a place, and is realized through the sacrifice of important attributes and values such as prosperity, freedom, or in the most extreme situation, life itself" (pg. 350). What Shamai is saying here is that when one operates on the greatest sense of place they willingly give up opportunities to prosper and be free to do whatever they want. Ultimately the person who cares about a place that much will forfeit other major life occurrences.

For example, if a person could obtain a higher paying job in another place, but that person chooses to remain in their current place, in a lower paying job, because they have a strong sense of place and are operating on a category 6 sense of place. Notice how this naturally encumbers individuals on the one hand and reduces their personal freedom.

Seamon's (1979) book "A Geography of the Lifeworld: Movement, Rest, and Encounter" echoed many of the same ideas that Relph and Tuan proposed, but put greater focus on the experience in place. Seamon's book explores what different environments mean to the individual and how they shape an individual's perspective and identity. Seamon argues there are three general patterns in environmental experience. The patterns are movement, rest and encounter. *Movement* can be described as everyday habitual body movements such as walking down the same side of a particular road. Repetitions of particular actions in particular contexts are described as "place ballets" (Seamon, 1979). *Rest* deals with rootedness and feeling at home in a place, echoing Relph's description. *Encounter* entails gaining greater awareness of a place when we carefully notice details of that place so that places standout and become salient things in an individual (Seamon, 1979).

Agnew (1987) contends that within social science place has been conceptualized using three major elements; locale, location, and sense of place. Locale contains the physical settings in which people form social relations. Location represents the geographical boundary which encompasses the social interaction as it is defined by social and economic processes operating on a larger scale, and sense of place encompasses the local structure of feeling. In Agnew's conceptualization sense of place becomes one part of a three-part framework for understanding, with sense of place containing primarily the affective associations. As Gustafson, (2001a) notes, for Agnew meaningful places come about in a social context happening through social relations, but those relations become bounded by geographic, economic, and cultural surroundings. This three-pronged framework will be echoed by later geographers and anthropologists.

Fishwick & Vining (1992) conducted a series of qualitative interviews in Illinois state parks to examine individuals' intentions in visiting recreation sites. Using content analysis of interviews, they gauged how past experiences influence individuals' reasons for visiting different sites. The results revealed that one, some individuals found meaning in specific qualities of a place, giving that place a sense of place all on its own; and two, past experiences played a significant role in whether an individual felt a sense of place, involving a combination of setting, landscape, ritual, routine, and how the new place related to the other places an individual had been. This suggests that sense of place is multi-faceted and places themselves carry meaning external to an individual's cognition of that place (Fishwick & Vining, 1992).

Robert Hay conducted one of the most often-cited studies of sense of place. He combined work from Relph (1976), Tuan (1974, 1980), and Hummon (1992), to create what he called a "rooted sense of place." Hay conceptualized sense of place on two dimensions. He agreed with Relph, Tuan, Buttimer (1980), and others that having a sense of place depended on having important or meaningful experiences in a place. He added a second dimension that emphasized how length of time in place was crucial to the development of a rooted sense of place (Hay 1998).

He argued that a person who dwells for a long period of time in a place developed a place perspective that allowed them to feel like an insider within a geographically- and culturally-bound place. To the degree that the individual has salient experiences in that place, they felt rooted in that place. Hay adds to Relph and Tuan, that individuals retain an ideal or general sense of place, which is strengthened and enhanced through the rooted sense of place. This general or ideal sense of place creates a place perspective that gives the individual a template allowing them to compare and contrast new places, a toolkit of sorts (1998).

Massey (1995) focuses more on the process of place-making rather than on what place means. Places are often depicted as objects containing their own perspective. These perspectives, based on history and tradition, become highly demarcated and are continually remade and rethought of, giving places an evolving nature rather than a static state (D. Massey, 1995; Massey, Allen, & Sarre, 1999; Massey, 1991). Individuals continually go through a process of place-making and they do this through social interactions, thus giving rise to a global sense of place. Some argue that Massey's sense of place is a more progressive sense of place, taking it further than Agnew (Gustafson, 2001a).

Phenomenologist, John Casey argued that place is both a general and specific perception that can affect an individual on multiple dimensions. Drawing on Heidegger's *dwelling* concept he argued that place comes before space, or, that humans live, work, play, etc. in places, not in empty spaces (Casey 1996). Other biotic and abiotic forms occupy the other spaces on Earth, thus giving meaning to space which is what makes it have places (Casey, 1996). Casey argued that perception provides the primary starting point for existing and living in place and space (1996), making it the primary starting point for when individuals develop a sense of place.

Specifically, Casey emphasized that perception frames sensory data, including data beyond that taken from the initial experience. He considered more embedded perceptions that happen in the initial place-self formation, and that convey meaning to place (Casey, 1996). Merleau-Ponty denotes this conveyance as depth, "a 'primordial depth' that, far from being imputed to sensations, already situates them [i.e. sensations] in a scene of which we ourselves form part" (Casey, 1996). Being surrounded by

18

depths and horizons that exist completely separate from ourselves, a person finds themselves in the midst of an "entire teeming place-world" rather than a confusing jumble of free-floating sensory data Casey (1996). In other words, places have phenomenological characteristics that have depth and provide realms upon which a person can attune their perception and form a strong sense of place. The phenomenological aspect of place happens because humans dwell in places. We are born into this world and keep returning to this world as already placed in the world (Casey, 1996) This means that all of our experiences happen in place(s) throughout our lifetimes, not in a sealed vacuum. Through our experiences happening in place we keep coming back to places or we discover new places. Either way, place goes where we go.

Uses in the Field of Anthropology

Anthropologists, whose main window into knowing the world is via culture, have several specific uses of 'place'. Even in the postmodern era when people lead more transient lives, "places continue to be important in the lives of many people, perhaps most, if we understand by place the experience of a particular location with some measure of groundedness (however unstable), sense of boundaries (however permeable), and connection to everyday life, even if its identity is constructed, traversed by power, and never fixed" (Escobar, 2001. pg. 140).

Anthropologists also consider place as an important element of collective myths, stories, daily actions, etc. of a community, and these meanings then inform an individual's knowledge, awareness, and values about place (Kahn, 1996). "Places capture the complex emotional, behavioral, and moral relationships between people and their territory" (Kahn 1996,pg. 168).

Basso (1996) also conceptualizes sense of place using Heidegger's notion of *dwelling*, which assigns importance to the forms of awareness or consciousness with which people perceive and understand geographical space. These forms could be homes, treehouses, grocery stores, office spaces,

19

etc. Regardless of whatever thing the form takes, dwelling consists in the multiple "lived relationships" that people maintain with places, and these "lived relationships" allow space to acquire meaning and have value in the consciousness of one's being (Basso, 1996). These multiple lived relationship includes features of the physical environment as well as the social environment (Cloke & Jones, 2001). Thus, through dwelling in places, individuals form strong connections with their environments and uses places to help shape their identity. As Basso (1996) writes, "selfhood and placehood are intertwined" (pg. 86).

Pred (1983) combines structuration processes with sense of place. For humanistic geographers "It [sense of place] is always an object for a subject. It is seen, for each individual, as a center of meanings, intentions, or felt values; a focus of emotional or sentimental attachment; a locality of felt significance" (pg. 49). Pred's combination allows sense of place to be modeled and shows how places themselves might be salient to an individual but also somewhat generalizable to groups or communities to which the individual belongs. This connects to Basso's (1996) claim that place is both universal and individual.

Hay (2009) also notes another research track on sense of place developed by structuralist researchers such as Agnew (1989) and Paasi (1991). Agnew (1989) and other structuralists further analyzed place in three constituent elements; *locale, location,* and *sense of place.* Locale was defined as the settings in which social relations become constituted. Location became the geographical setting for the social interactions, and sense of place became the common frame of experiences. Furthermore, these different structures or domains of place operate on different scales so that a local place is in itself part of a regional dynamic system (Paasi, 1991). Structuralists viewed culture as behaving similarly to local places, in that local cultures become part of a regional, dynamic system (Paasi, 1991).

For many researchers in place studies, culture and place are intertwined. That is local, regional, and national cultures are contextualized by place, by the physical places where-in the culture occurs

20

(Basso, 1996; Casey, 1996; Escobar, 2001b; Gustafson, 2001b). Places gives culture meaning and vice versa (Escobar, 2001a). As this paper progresses this reciprocity will become self-evident.

Huff (2006) examined sense of place as it related to an indigenous people rather than the contemporary society. In a study examining maize, story-telling, and the Mayan sense of place, Huff reviewed oral stories to find out why maize is sacred to the Mayans. For the Mayans, maize is not just a food stuff, but rather a sacred item that encapsulates their connection to the land, or to their place. As Huff writes, "Maize is not solely sustenance for the Maya peoples; it is linked to key spiritual, cultural, and social activities- the all encompassing "substance of life" (Huff, 2006 pg. 81).

Huff argues that one way researchers can understand the cultural meanings of these all encompassing things such as maize, especially for oral traditions, is to try and locate a people's sense of place. For the Huff, understanding the Mayan's sense of place helped her understand why maize became so sacred. It had much to do with the role of maize in the creation of the people. The creation myths reveal that the early Mayan agriculturist infused maize with many symbols such as the balance of life and death, subsistence and nourishment, and balanced movement of the cosmos (Huff, 2006). To this day Mayans will plant ears of corn in their own home and not cultivate them. They will let them grow in as natural a life cycle as possible and this represents the individuals life cycle (Huff, 2006). Though this article was published in a geographical publication, it reads more like an anthropological study.

Though Stewart (1996) doesn't examine an indigenous culture in the classical sense she does examine a more rural landscape that hearkens back to past. Stewart examined the "hills and hollars" of West Virginia to see how generations of families have come to form strong identities with rural West Virginia. She focuses greatly on the local, writing "far from being a timeless or out-of-the-way-place, the local finds itself reeling in the wake of every move and maneuver of the center of things" (pg. 137). She examined local landscapes and the people in them for their language, cultural meanings, values,

21

and family bonds. Stewart concluded that in the hills and hollars their resides rich cultural heritage that has become tied to the land.

The local communities have been faced with hard times. Perhaps they could find better lives somewhere else, but they choose to stay; their lives are tied to the land (Stewart, 1996). This connects to Basso's idea about placehood and selfhood being intertwined, suggesting that once an individual forms a definite sense of place that entails rootedness and dwelling than it would become very difficult for that individual to leave that place despite the harshness of the life.

Stewart's (1996) study also shows how the local aspect can enrich a sense of place. She writes, "the sense of place grows dense with social imaginary- a fabulation of place contingent on precise modes of sociality and tactile, sensate force" (pg. 137). Sense of place finds more precision on a local scale, therefore it grows in character and possibly meaning.

A more recent exposition of studies shows a wide variety of different usages of sense of place, and therefore different meanings. In this next section I will discuss several studies that can best be categorized as miscellaneous. These studies go no further back than 2000.

A More Recent Review

Klanicka, Buchecker, Hunziker, & Müller-Böker (2006) looked at two different interest groups and how they formed a sense of place centered around the Swiss Alps. They examined tourists' and locals' sense of place and tried to answer the basic question of do outsiders and insiders relate to the same place differently. The researchers conducted semi-structured interviews with 23 individuals. After conducting and analyzing the interviews they created a triangular sense of place model with several categories clustered around 3 distinct poles; individual, society, and existence. Individual represented the very subjective needs of a person. Society represented the needs of social integration, and existence represented people's livelihoods and property.

They found that the landscape became relevant to both tourists and locals, but for different

22

reasons. They also found that social relationships had significance in the formation of a sense of place for both locals and tourists, as well as the local culture. Overall, the researchers found that both locals and tourists possess a similarly differentiated sense of place. Many of the same key categories have significance for both groups. The distribution of key concepts proved to be the major difference. Locals found significance in categories that represented everyday life; occupation, property, and social relationships, which together formed their affiliation. The tourists shape their sense of place primarily through the aesthetics and characteristics of the landscape.

Nanzer (2004) contributed to this issue in his study on residents of Michigan. He followed Jorgenson & Stedman's (2001) lead and assessed place identity, place dependence, and place attachment on a statewide scale. He surveyed a large sample of Michigan residents to see if they had a sense of place for the state, thus increasing the spatial scale. They found that overall 80.1% of the respondents indicated a sense of place for the state of Michigan. This supports Low and Altman's (1992) and Kaltenborn & Bjerke's (2002) claim that sense of place can be formed for a place of any size (Nanzer, 2004).

A study by Windsor & Mcvey (2005) examined what happens to people's sense of place if place becomes threatened and how do they react? They looked at what happened to an indigenous people's sense of place when their sacred lands were lost. They examined what happened when, in the 1940's, the Canadian government allowed a dam building project on the Nechako River, despoiling the lands of the Cheslatta T'En Canadian First Nation. Known as the Kemano Dam, the project flooded the river valley, including several Cheslatta T'En reservations and graveyards where the Cheslatta T'En had been for over 10,000 years, subsisting by hunting and gathering (Windsor & Mcvey, 2005). After displacement the destruction of graveyards was particularly devastating, indicated by the 50% rise in deaths from suicide or alcohol and drug abuse by 1990 (Windsor & Mcvey, 2005). Losing the land brought about a group-wide identity crisis, leading to a host of social issues (Windsor & Mcvey 2005).

23

One of the major themes of sense of place, which will be discussed further, has to do with its integrative nature. Sense of place as a concept and through its usage in the different disciplines naturally cuts across many dimensions and in turn, several disciplines. So, newer research would inevitably look at sense of place in an integrative perspective, rather than just within a discipline. That is, sense of place would be defined or understood as a concept that attempts to integrate its anthropological, geographical, and psychological aspects.

Integrative Perspectives

The literature review of sense of place also shows studies that take an integrative approach, with sense of place being a major concept in their framework. Mitchell Thomashow discusses how individuals can and do shape their perspective to become more ecologically grounded. Much of his work takes the ideas of Proshansky, Relph, and Tuan and ties them together.

Thomashow (1995) defines ecological identity as all the various ways individuals construe themselves in relationship to the earth manifested in a person's personality, values, actions, and their sense of self. Like Proshansky, Thomashow (1995) argues that nature becomes an object of identification and that not only do people carve out their identities through cultural and social interactions, but also a person's relationship to the earth, their perception of the ecosystem, and direct experience of nature. Also similar to Proshansky, Thomashow places importance on how sensing places shapes or influences one's ecological identity. Thomashow looked at how landscapes, physical spaces, or local places become a mediator for identity development.

One of his starting points revolves around childhood memories of special places. In recounting ecological identity work among individuals from different cultures, Thomashow (1995) shows how a similar pattern emerges. He writes, "despite the great variety of international and cultural experience, there is a striking thematic pattern: whether the person is from an Asian tropical rain forest, an African savanna, a Latin-American city, a European valley, or a North American farm, they tell a similar story.

24

They all have fond memories of a special childhood place, formed through their connections to the earth via some kind of emotional experience, the basis of their bonding with the land or neighborhood" (pg. 9). Usually, the memories center around play experiences, involving exploration, discovery, adventure, imagination, and independence (Thomashow, 1995). Thomashow's writing about memories of special childhood places has partly informed my interview guide for my semi-structured interviews. The first section of the guide is dedicated to fleshing out an adults' memory of special childhood places.

During the period of middle childhood (ages 9-12) children undergo a process of *place-making* wherein children use the local environment to expand their sense of self. In these formative years children leave the safety of the home and explore their immediate surroundings. In doing so their perceptions of the immediate environment undergo a transformation and the child realizes that he or she has a unique perception of the world (Thomashow, 1995). More importantly for this study this place-making period allows the child to form strong bonds with places, landscape types, ecosystems, etc. which later lead to an adult's sense of place (Thomashow, 1995).

Thomashow (1995) notes that sense of place literally entails the roots of ecological identity. He writes, "ideas such as bioregionalism, sustainability, material simplicity, community, citizenship, decentralization, environmental psychology, and others were integrated in this one expression" (pg. 192-193). The statement examines sense of place from a holistic perspective.

Thomashow (1995) also notes that by discovering the place where you live or your immediate environment (i.e. the local) and seeing the patterns in the landscape, experiencing the changes in the ecosystem, and how humans live off the land and share the land with other species, one can directly expand their circle of identification to a more biocentric or ecocentric perspective, which harkens back to Leopold's call in his land ethic. Leopold wrote that we can be ethical only in regards to something

25

we can see, feel, understand, love, or otherwise have faith in. In other words, we can be ethical in regards to something we can sense.

It seems to me then that forming an ecological identity or conscious which might actually lead to practicing a more sustainable lifestyle or acting more ethically towards nature depends greatly upon one learning more about one's sense of place and how one relates to the greater biotic community in which he or she lives. Thomashow (1995) notes work by Daniel Kemmis, who argues that people should "reinhabit" their community landscape and transcend the individualistic notions that drive them apart, and by extension drive them away from the natural environment in which they live. Accordingly, Kemmis is lead to work by Gary Snyder and Wendell Barry who show how people who coexist in a habitat form common bonds through the process of working together (Thomashow, 1995).

Another theme from Thomashow and others deals with bioregionalism. Thomashow (1995) argues that bioregionalism uses the affective and cognitive ties to local landscapes stemming from people's sense of place as a way to build cultural, political, and economic arrangement at a more naturally occurring regional level. Bioregionalism as a movement seeks to have residents of ecological, social, and culturally distinct regions live more harmoniously with their local landscapes, in short, to become one with the nature of place (Thayer, 2003).

Thayer (2003) articulates his theory of bioregionalism as a lifeplace, arguing that the lifeplace concept seeks to emplace biophilia on a physical and social scale and suggests a means of living in place through a stronger understanding of, respect for, and care of a naturally bounded region or territory. The bioregion is seen as the most logical locus and scale for a sustainable and regenerative community to take place. Bioregionalism or the lifeplace concept literally grounds people in their local place through matching their sense of place with the sense of place of the region, thereby providing a tighter coupling between humanity and place (Thayer, 2003).

26

Thayer (2003) writes that "people who stay in place may come to know that place more deeply. People who know a place may come to care about it more deeply. People who care about a place are more likely to take care of it" (pgs. 5-6). Though the axioms imply a certain causal flow there is a unifying theme in these axioms, which is sense of place. The axioms have helped create a structure for how sense of place works on a more dynamic level. Sense of place can go beyond mere factual or academic knowledge of a place or landscape. Sense of place can work to incorporate peoples' compassion and love of place, which would provide even stronger connections to places. Forming a sense of place would literally provide the roots that Thomashow mentioned above.

In summation a review of the literature reveals two major tracks with one track led by phenomenologists, geographers, and anthropologists and the other track led by psychologists. The two tracks complement each other and though sense of place has been conceptualized in many different ways there seems to be a basic premise and follow a couple of themes. In the following chapter I will discuss this premise and the themes. The question becomes, can sense of place as it has been conceptualized be used in the modern sustainability movement?

Chapter 3

Discussion

Sense of place has clearly evolved over time as it has been used in the different disciplines. Though the foundational aspects of the concept have remained relatively the same, the different disciplines have added different perspectives, giving sense of place a slightly different twist for each discipline. However, it is clear that one could categorize the human geographers, phenomenologists, and anthropologist together in having the same foundational principles. Psychologists have different foundational principles which leads to different dynamics.

The early human geographers, i.e. Relph, Tuan, and Buttimer, based their ideas about sense of place on both an experiential perspective as well as Heidegger's *dwelling perspective*. Using the dwelling idea, the thought was that individuals can't lay down roots in a place or identify with a place or call a place a home until they began to see that they dwell in a place. In this sense place becomes more than a location where a person just works or vacations or visits family. Place becomes something of great meaning and importance, and has an ability to dramatically shape an individual in various ways (Wollan, 2003).

Studies since Relph, Tuan, and Buttimer have done the same. Basso, Feldman, Casey, Hay, and Shamai referenced and built upon the same ideas of Relph, Tuan, and Buttimer. However, from there they each took sense of place to a slightly different perspective.

The human geographers such as Hay began to view sense of place as multilayered, consisting of both a general sense of place or an idealized sense of place and a more specific sense of place that

28

related an actual place. This allows for individuals the capacity to try and fit into new places (Hay, 1998b). I rather like Hay's work because it also drew on Heidegger's work like Relph and Tuan, but also made sense of place more fundamentally sound. Hay's concept allows for individuals to take their sense of place with them to new places and juxtapose that or overlay that with the new place, thus, allowing sense of place to change and evolve as the individual changes and evolves.

I think the contributions of the human geographers, along with the phenomenologists, have made sense of place more dynamic because in their work there allows for more reciprocity between people and places. This will impact the sustainability movement, which I will discuss later. This reciprocity also gives more value to places themselves.

For the human geographers places themselves can take on their own perspectives or individual meanings. Whereas psychologists have concerned themselves primarily with how a person has been shaped by place, human geographers concern themselves with how places have been shaped by humans. In their perspective, along with phenomenologists, people learn from being emplaced, but places themselves change and their meanings change from people being emplaced.

Psychologists, beginning with Proshansky and continuing up to the present, only examine the different psychological aspects of people and places. In this way, I think that sense of place, as understood by human geographers, phenomenologists, and to some degree, anthropologists, is a much more dynamic integrative concept. Yes, it is harder to define and empirically assess this way, but it has more depth as a concept. To approach it, one needs to think integratively.

That is exactly what Thomashow has done in his work. He integrated the experiential underpinnings developed by the early geographers and phenomenologists combined that with the meanings individuals gain from having a sense of place (both affective and cognitive, social and individual) and fused that with ecological science. Thomashow understood that people change over time and since people can and do change over time, their understanding and perceptions of place change over time. For him, this means that place can have large impacts into how people view their relationships, both to each other and to other places. Thomashow hit on all the major themes, which I will discuss next.

An analysis of the literature review reveals three themes. For one thing, sense of place seems to encapsulate mainly the affective bonds individuals have with places while place attachment or place identity captures the cognitive bonds individuals have. Secondly, no matter what track a researcher pursues with place studies, meaningful experiences become important for an individual to form a sense of place or develop a place identity. Finally, researchers discuss sense of place more in terms of the local versus the regional, national, or global scales.

Sense of place primarily entails affective bonds with places. Both Relph and Tuan talked about how we build meaningful connections with place through how we come to feel about places. We find great meaning in certain places because we tend to have positive feelings in those places. We feel a sense of serenity, peace, joy, happiness, etc.

For Relph rootedness and insideness can only happen in places that make us feel good. Tuan took the affective bonds further in topophilia, where he talked about a deep love of place and topography. Whether it is a mountain lake or a bustling downtown the place makes us feel something very positive. These good feelings tend to happen because of the second theme, that of salient experiences.

Having a sense of place depends greatly upon having meaningful experience in place. A person needs to have some important experience in the place that forms the object of their sense of place. Repeated experiences in that places seems to reinforce one's sense of place, but extended periods of time doesn't necessarily guarantee that a person will form a sense of place for that place.

In sum, based on this review of the literature I have made to date, I conclude that it not possible to create a unifying concept for sense of place. On the one hand the concept of sense of place has many

30

dynamics and it is highly variable depending on the discipline and/or context in which it is used. In one analytical framework sense of place can be based on cognitive connections whereas in another sense of place is based on using cognitive, affective, and behavioral dimensions. This is not to be interpreted as a failure by any means but rather an expression of the term's richness since, based on the literature I have engaged, creating a unifying concept for sense of place would in many ways diminish its effectiveness in these different disciplinary and analytical contexts.

Chapter 4

Conclusion

Although my literature review and analysis argues against a unifying concept of sense of place, I nevertheless want to promote the concept of sense of place, in its various forms, could be used to communicate and bring about a societal-wide shift in perspective. Raising awareness of individuals and communities about sense of place could help to localize those individuals and communities so that, even in a more globalized economy, place becomes valued and important. Yet, there seems to be an asymmetry in the debate between global and local on how the global economy has changed cultures (Escobar, 2001b). Place brings the local back into focus. As Escobar (2001) notes global is often equated with space, capital, history, and agency whereas the local is equated with place, labor, and tradition. As a result, the more societies become global in scale the less valuable local traditions, cultures, labor forces, languages, environments and other aspects of life become (Escobar, 2001). Escobar (2001) and others argue that it may be time to reverse some of this direction of development by focusing anew on the vital role place and place-making (which starts from place sensing) has in the formation of culture, nature, and economy.

Though sense of place could localize the perspectives of individuals, I cannot say with any confidence that researchers can find a unifying theme. Sense of place has evolved over time, and the different disciplines have some similarities in the properties of sense of place for how they use it, but the foundational aspects of sense of place (i.e. the being in place promoted by the geographers and phenomenologists and the identity development promoted by the psychologists) counteract each other, or at least, they do not mix well together. The different meanings preclude a unifying theme. So, to conclude and answer the question of can there be a unifying theme, the answer is no, until the disciplines change the foundational principles, or agree on the foundational principles.

What I mean is that we as a society, rather than I as an individual, need to collectively think of place as something more than just one thing, as more than just "it's a place where I work" or "it's where I go to vacation." I don't know if that can happen in a western reductionist society. Basso, and others including Gary Snyder and Wendell Berry, have discussed this point. If society can start to shift the collective perspective on place from a single thing to a landscape that integrates many aspects, than perhaps individuals within society can start to agree on what sense of place means and it would have potential to be used in the sustainability movement.

Sustainability calls for individuals and society to be more aware of how we impact places through our daily actions. Instead of becoming more aware, one could argue that we should enlarge our sense of things, and since sustainability focuses on environments or places, then it becomes natural to think that changing one's sense of place could potentially work on how they view their relationship to places. Gary Snyder calls for this very thing (Benesch & Schmidt, 2005). Snyder writes, "it is not enough just to 'love nature'…Our relation to the natural world takes place in a *place*, and it must be grounded in information and experience" (Snyder, 2010 pg. 39).

Sense of place could be used in sustainability if it becomes a more unified concept based on integrative approaches. For, sense of place potentially entails both specific and general qualities of

33

place; it entails cognitive, affective, and geographical ties, and it potentially relates both to individuals and to places themselves. The crux of the issue isn't whether people have a sense of place; it is whether people know they have a sense of place and if they do have one, what dimensions of place does that entail I think if people were given an opportunity to reflect on their sense of place, as Thomashow suggests, they would certainly become more aware of their own environmental ethical code and it would at least open the door to the possibility of them become more biocentric or ecocentric. That is partly what drives work by researchers such as Thomashow and Thayer.

In conclusion, sense of place, coined by Relph and developed throughout several different disciplines, could become a concept that is integrative and unifying. Sense of place, could potentially be used to work towards sustainability if people begin to see places differently, more dynamically. I think that is at the heart of the crossroads discussed by Devine-Wright & Clayton.

Future Directions

Places entail so much more than we originally think or envision. They are dynamic, evolving, and ever-changing. They are not frozen in time. Our sense of that needs to evolve as well. In other words, it is not what we think a place is, it is how we perceive it to be. For sense of place to play an active role in societal change, it must be viewed as a process that happens over time or experientially rather than an outcome. In other words, sense of place studies should incorporate the process of place-making as well as the object of the place-making. To fully capture and take advantage of the richness of sense of place, researchers should engage interdisciplinary approaches and employ mixed-methods that integrate the concept's many dimensions, similar to Thomashow.

References

Agnew, J. A. (1987). *Place and politics: The geographical mediation of state and society*. Allen & Unwin Boston. Retrieved from http://www.getcited.org/pub/102554133

Agnew, J. A. (1989). The Devaluation of Place in Social Science. In J. A. Agnew & J. S. Duncan (Eds.), *The Power of Place: Bringing Together Geographical and Sociological Imaginations* (pp. 9–29). Boston: Unwin Hyman.

Basso, K. H. (1996). Wisdom sits in places: notes on a western apache landscape. In Feld & Basso (Eds.), *Senses of Place* (pp.53-90). Santa Fe, NM: School of American Research Press.

Benesch, K., & Schmidt, K. (2005). *Space in America: Theory, History, Culture*. Rodopi.

Buttimer, A. (1980). Home, reach, and the sense of place. *The Human Experience of Space and Place*, *166*, 187.

Casey, E. S. (1996). How to get from space to place in a fairly short stretch of time. In Feld & Basso (Eds.), *Senses of Place* (pp. 13-52). Santa Fe, NM: School of American Research Press.

Cloke, P., & Jones, O. (2001). Dwelling, place, and landscape: an orchard in Somerset. *Environment and Planning A*, *33*(4), 649 – 666. doi:10.1068/a3383

Cuba, L., & Hummon, D. M. (1993). A Place to Call Home: Identification with Dwelling, Community, and Region. *The Sociological Quarterly*, *34*(1), 111–131.

Devine-Wright, P., & Clayton, S. (2010a). Introduction to the special issue: Place, identity and environmental behaviour. *Journal of Environmental Psychology*, *30*(3), 267–270. doi:10.1016/S0272-4944(10)00078-2

35

Devine-Wright, P., & Clayton, S. (2010b). Introduction to the special issue: Place, identity and environmental behaviour. *Journal of Environmental Psychology, 30*(3), 267–270. doi:10.1016/S0272-4944(10)00078-2

Escobar, A. (2001a). Culture sits in places: reflections on globalism and subaltern strategies of localization. *Political Geography, 20*(2), 139–174. doi:10.1016/S0962-6298(00)00064-0

Escobar, A. (2001b). Culture sits in places: reflections on globalism and subaltern strategies of localization. *Political Geography, 20*(2), 139–174. doi:10.1016/S0962-6298(00)00064-0

Feld, S., & Basso, K. H. (1997). *Senses of Place*. Santa Fe, NM: School of American Research Press.

Feldman, R. M. (1990). Settlement-Identity. *Environment and Behavior, 22*(2), 183 –229. doi:10.1177/0013916590222002

Fishwick, L., & Vining, J. (1992). Toward a phenomenology of recreation place. *Journal of Environmental Psychology, 12*(1), 57–63. doi:10.1016/S0272-4944(05)80297-X

Gruenewald, D. A. (2003). Foundations of place: A multidisciplinary framework for place-conscious education. *American Educational Research Journal, 40*(3), 619.

Gustafson, P. (2001a). Meanings of Place: Everyday Experience and Theoretical Conceptualizations. *Journal of Environmental Psychology, 21*(1), 5–16. doi:10.1006/jevp.2000.0185

Gustafson, P. (2001b). Meanings of Place: Everyday Experience and Theoretical Conceptualizations. *Journal of Environmental Psychology, 21*(1), 5–16. doi:10.1006/jevp.2000.0185

Hay, R. (1998a). A Rooted Sense of Place in Cross-Cultural Perspective. *Canadian Geographer, 42*(3), 245–266. doi:10.1111/j.1541-0064.1998.tb01894.x

Hay, R. (1998b). Sense of place in a developmental context. *Journal of Environmental Psychology, 18*(1), 5–29. doi:10.1006/jevp.1997.0060

Heidegger, M. (1996). *Being and Time: A Translation of Sein und Zeit*. SUNY Press.

Hidalgo, M. C., & Hernandez, B. (2001). Place attachment: Conceptual and empirical questions. *Journal of Environmental Psychology, 21*(3), 273–281. doi:10.1006/jevp.2001.0221

Huff, L. A. (2006). Sacred Sustenance: Maize, Storytelling, and a Maya Sense of Place. *Journal of Latin American Geography, 5*(1), 79–96.

Hummon, D. M. (1992). Community Attachment: Local Sentiment and Sense of Place. In I. Altman & S. M. Low (Eds.), *Place Attachment*. New York: Plenum Press.

Jorgenson, B., & Stedman, R. (2001). Sense of place as an attitude: lakeshore owners attitudes toward their properties. *Journal of Environmental Psychology, 21*(3), 233–248. doi:10.1006/jevp.2001.0226

Kahn, M. (1996). Your Place and Mine. In S. Feld & K. Basso (Eds.), *Senses of Place*. Santa Fe, NM: School of American Research Press.

Kaltenborn, B. P., & Bjerke, T. (2002). Associations between environmental value orientations and landscape preferences. *Landscape and Urban Planning, 59*(1), 1–11. doi:10.1016/S0169-2046(01)00243-2

Klanicka, S., Buchecker, M., Hunziker, M., & Müller-Böker, U. (2006). Locals' and Tourists' Sense of Place: A Case Study of a Swiss Alpine Village. *Mountain Research and Development, 26*(1), 55–63.

Korpela, K. M. (1989). Place-identity as a product of environmental self-regulation. *Journal of Environmental Psychology, 9*(3), 241–256. doi:10.1016/S0272-4944(89)80038-6

Low, S. M., & Altman, I. (1992). Place attachment: A conceptual inquiry. *Place Attachment*, 1–12.

Massey, D. (1991). A global sense of place. *Marxism Today, 35*(6), 24–29.

Massey, D. (1995). The conceptualization of place. *A Place in the World, 4*, 45–77.

Massey, D. B., Allen, J., & Sarre, P. (1999). *Human Geography Today*. London, UK: Polity Books.

Moore, R. L., & Graefe, A. R. (1994). Attachments to recreation settings: The case of rail-trail users. *Leisure Sciences, 16*(1), 17–31. doi:10.1080/01490409409513214

Nanzer, B. (2004). Measuring Sense of Place: A Scale for Michigan. *Administrative Theory & Praxis, 26*(3), 362–382.

Paasi, A. (1991). Deconstructing Regions: Notes on the Scales of Spatial Life. *Environment and Planning, 23*, 239–256.

Pred, A. (1983). Structuration and Place: On the Becoming of Sense of Place and Structure of Feeling*. *Journal for the Theory of Social Behaviour, 13*(1), 45–68. doi:10.1111/j.1468-5914.1983.tb00461.x

Prince, H. (1974). Sense of Place. *Area, 6*(1), 71–72.

Proshansky, H., Fabian, A., & Kaminoff, R. (1983). Place-identity: Physical world socialization of the self. *Journal of Environmental Psychology, 3*(1), 57–83.

Proshansky, H. M. (1978). The City and Self-Identity. *Environment and Behavior, 10*(2), 147 –169. doi:10.1177/0013916578102002

Relph, E. C. (1976). *Place and Placelessness*. London: Pion Press.

Scullin, J. (2009, March 1). Local Markets, Co-Ops Reflect a Growing Trend. *The Tampa Tribune*.

Seamon, D. (1979). *A Geography of the Lifeworld: Movement, Rest and Encounter*. Croom Helm.

Seamon, D. (1996). A Singular Impact: Edward Relph's Place and Placelessness. *Environmental and Architectural Phenomenology Newsletter, 7*(3), 5–8.

Seamon, D. (2000). A Way of Seeing People and Place: Phenomenology in Environment-Behavior Research. In S. Wapner, J. Demick, T. Yamamoto, & H. Minami (Eds.), *Theorectical Perspectives in Environment-Behavior Research*. New York: Plenum Press.

Seamon, D., & Sowers, J. (2008). Place and Placelessness, Edward Relph. In P. Hubbard, R. Kitchen, & G. Vallentine (Eds.), *Key Texts in Human Geography*. London: Sage Publications.

Semken, S., & Freeman, C. B. (2008). Sense of Place in the Practice and Assessment of Place-Based Science Teaching. *Science Education, 92*(6), 1042–1057.

Shamai, S. (1991). Sense of place: an empirical measurement. *Geoforum, 22*(3), 347–358. doi:10.1016/0016-7185(91)90017-K

Snyder, G. (2010). *The Practice of the Wild*. Counterpoint Press.

Stedman, R. C. (2002). Toward a Social Psychology of Place: Predicting Behavior from Place-Based Cognitions, Attitude, and Identity. *Environment and Behavior*, *34*(5), 561–581. doi:10.1177/0013916502034005001

Stewart, K. (1996). An occupied place. In K. Basso & S. Feld (Eds.), *Senses of Place*. Santa Fe, NM: School of American Research Press.

Thayer, R. L. (2003). *LifePlace: Bioregional Thought and Practice* (1st ed.). Berkeley: University of California Press.

Thomashow, M. (1995). *Ecological Identity: Becoming a Reflective Environmentalist*. Cambridge, MA: The MIT Press.

Tuan, Y.-F. (1974). *Topophilia: A Study of Environmental Perception, Attitudes, and Values*. New York: Columbia University Press.

Williams, D. R., Patterson, M. E., Roggenbuck, J. W., & Watson, A. E. (1992). Beyond the commodity metaphor: Examining emotional and symbolic attachment to place. *Leisure Sciences*, *14*(1), 29–46. doi:10.1080/01490409209513155

Williams, D. R., & Roggenbuck, J. W. (1989). Measuring place attachment: Some preliminary results. In *Symposium on Outdoor Recreation Planning and Management, National Recreation and Park Association Research Symposium on Leisure Research*. San Antonio, USA.

Windsor, J. E., & Mcvey, J. A. (2005). Annihilation of Both Place and Sense of Place: The Experience of the Cheslatta T'En Canadian First Nation within the Context of Large-Scale Environmental Projects. *The Geographical Journal*, *171*(2), 146–165.

Wollan, G. (2003). Heidegger's philosophy of space and place. *Norsk Geografisk Tidsskrift*, *57*(1), 31–39. doi:10.1080/00291950310000802

Biography

Charles R. Milling grew up in Texas. He attended Texas Tech University, where he received his Bachelor of Arts in English Literature in 2001. He went on to receive his Masters of Education in Higher Education Administration in 2005. He has taken on an administrative role at South College-Asheville in Asheville, NC.

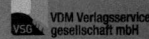